Why Do - Design Optimisation?

Author:

David Spicer

ACKNOWLEDGEMENT

I wish to take this opportunity as Chairman of NAFEMS Education and Training Working Group to thank the Working Group members for their help and support in the preparation of this booklet. The composition of the Working Group is:

Adib Becker	University of Nottingham
Steve Dalton	Consultant
David Ellis	IDAC
Steve Hardy	University of Wales, Swansea
Trevor Hellen	Consultant
Bob Johnson	DAMT Ltd
Tom Kenny	NAFEMS
Tony Luxmoore	University of Wales, Swansea
Anup Puri	BAE Systems
John Smart	University of Manchester
Bryan Spooner	Consultant
Jim Wood	University of Strathclyde

I would also like to thank Royston Jones for his efforts in his capacity as chief reviewer of this booklet.

John Smart
Chairman

Disclaimer

Whilst this publication has been carefully written and subject to peer group review, it is the reader's responsibility to take all necessary steps to ensure that the assumptions and results from any finite element analysis which is made as a result of reading this document are correct. Neither NAFEMS nor the authors can accept any liability for incorrect analyses.

Preface

NAFEMS is a non-profit making association of organisations using, developing, or teaching the various forms of numerical analysis in common practice.

This booklet is a continuation of the "How and Why .." set of booklets published by NAFEMS, designed to guide new and experienced analysts in a range of problem types. The booklets are written to introduce various analysis methodologies to both engineering managers and engineers, in a straightforward and informed manner. They are complemented by more detailed publications from NAFEMS; for example, the Introduction to Non-Linear Finite Element Analysis and Understanding Non-Linear Finite Element Analysis through Illustrative Benchmarks.

Optimisation has been used in industrial design for several decades. In particular, structural optimisation has made a major impact. Many of the general optimisation problems addressed in the past have been of a tactical nature rather than strategic. However, in recent years computing power and analysis capabilities have developed to the point where designs can now be assessed more comprehensively than ever before, with optimisation as an integral part of the design cycle. Moreover there is increasing industrial pressure to provide products whose cost, performance, durability and customer appeal have been refined and balanced to maximise market return.

In the past decade a number of research projects have been initiated. This booklet draws on experience from such research, particularly the FRONTIER project from the E.C.'s Esprit IV programme.

It aims to give an analyst who is new to this area an introductory guide to optimisation, the type of problems that have been satisfactorily solved, the possible pitfalls and how to proceed. It also provides details as to where further information can be found.

Contents

1 Introduction

'Optimisation' is about selecting the best option from a range of possible choices. It is natural to consider this when designing a new product. As the old saying goes, 'A job worth doing is worth doing well'. Why not produce a design which is the best in its class, if you have the means to do so? It may not cost any more in time and money to design and produce; and it might cost less. The customers for your product will be more pleased with it. Everyone will be happier, except your competitors.

But how to make the selection, that is the question. Maybe in practice, you would settle for something which is some way short of 'optimal'. Perhaps you would be happy with a significant improvement on an existing design. It amounts to a very similar problem; how to search for better designs, in a systematic, practical and affordable way.

In looking at design optimisation, we need to address some fundamental issues.

What are our objectives in designing a product? How 'well defined' does a design have to be in order to be optimisable? Given that a design evolves in a series of phases, where does optimisation fit in? Can it address the important high-value design decisions? What are the industrial process issues and benefits associated with putting design optimisation into practice? What are the benefits to the product?

We shall consider these questions in the sections which follow.

WHY DO DESIGN OPTIMISATION?

2 The Role And Objectives Of Design

We consider design in the context of a project whose goal is to produce a product according to some mix of criteria on cost, quality and timescale. Either a project which is being run and managed by an organisation whose production and sales support company goals concerned with growth and profitability. Or a project belonging to a non-profit making organisation such as a F1 Grand Prix team, more concerned with winning than with contributing to a company balance sheet.

What is the role of design in contributing to the success of a project or a company?

The creation of new products is a central feature of life in a modern society. The word 'product' is nowadays applied to a wide range of items, from household goods to holidays and financial packages.

In the field which the stock market calls 'Engineering', products large and small are created and sold. They underpin the plans of their producer companies, which in general are to try to remain in business, grow the business, and make money for shareholders and employees.

We all know products which have been around for decades. In a changing world, it is reassuring to know that you can still buy the same chocolate bar you enjoyed when you were young. Such products survive despite the competition, though they still need the public to be reminded occasionally of their merits, through marketing and advertising.

But products like these are the exception rather than the rule. To expand a business, indeed even to survive, new products need to be created from time to time, and turned into successful income generators. Even established products may well need to be manufactured in different ways, or out of different materials.

For the last two decades, the work of Professor Michael E. Porter [1],[2],[3] on competitive strategy has provided a definitive reference point on the nature of industrial competition. Porter identifies the 5 main influences on industry growth as

1. demographics
2. trends in needs
3. change in the relative position of substitutes
4. changes in the position of complementary products
5. penetration of the customer groups

WHY DO DESIGN OPTIMISATION?

In relation to product design, to quote Porter, "Product innovation can allow it to serve new needs, can improve the industry's position vis-a-vis substitutes, and can eliminate or reduce the necessity of scarce or costly complementary products. Thus product innovation can improve an industry's circumstances relative to the 5 external causes of growth, and thereby increase the industry's growth rate. Product innovations have played a major part in fuelling the rapid growth of motorcycles, bicycles, and chain saws, for example".

Porter reassures us that product innovation through outstanding design, achieved through design optimisation, can play a key role in assuring the future success of our businesses.

Competitive strategy has been further analysed by Hamel [4]. His work emphasises the importance of being 'driven by vision', more than being 'driven by competitors'. He stresses the need for companies to create fundamentally new 'competitive space', by being different rather than by being better or smaller. Hamel's thinking should encourage anyone carrying out design optimisation to adopt as wide a family of design candidates as possible, to maximise the opportunity to discover novel solutions.

Innovation can make all the difference between total success and complete failure.

An often quoted example is Team New Zealand.

In the 1995 America's Cup yacht race, Team New Zealand's boat Black Magic won 43 out of the 44 races. The race series is held over a period of around 3 months. The following is a quote from [5]. "One brilliant example of the power of short design cycle time was the extraordinary domination of the last Americas Cup yacht race by Team New Zealand and their yacht named Black Magic. Their performance in winning 43 out of 44 races of these very high technology machines is totally unprecedented. In describing how they were able to accomplish that magnificent achievement, the members of Team New Zealand provided the following statements, and many more:

1. 'the secret of Black Magic's awesome speed on the race course was the equally amazing speed of Team New Zealand's design cycle'
2. 'I can see the analysis on the computer screen, and within hours test those results on the water and feed observations to the designers'
3. 'By reinventing its design process, Team New Zealand was able to experiment with thousands of design options, and implement the best designs literally overnight'

The rate of learning that they were able to establish by means of very fast design and testing processes greatly exceeded that of their competitors, and they ended up dominating the competition."

WHY DO DESIGN OPTIMISATION?

3 The Design Cycle

3.1 What Is A Design?

The biggest impact is usually made by the fundamental design decisions arrived at early on. But early in the design process we may only have an outline of the final product. The requirements to be satisfied may still be somewhat unclear. Yet selections from various design options have to be made.

So how well defined does a design have to be before a rational selection process can be possible? What set of information is sufficient to qualify to be called a design?

Dictionary definitions of 'design' include

1. 'a plan or pattern from which a picture, building or machine may be made'
2. 'a preliminary sketch or outline (as a drawing on paper or a modelling in clay) showing the main features of something to be executed'
3. 'the process of selecting the means and contriving the elements, steps, and procedures for producing that will adequately satisfy some need'

For optimisation purposes, a design has to be viewed in relation to its ability to undergo evaluation of its merits. It is thus assumed to be a definition of a product or process which

1. is complete enough to allow the significant measures of its business effectiveness to be evaluated (preferably objectively and consistently via a clearly specified evaluation process)
2. is one candidate from a well defined class of designs, which allows 'cause and effect' to be established between a design change and the design's business effectiveness.

3.2 Design Phases: How Do The Concepts And Details Evolve?

The figure below shows an overview of a general design cycle.

Initial conceptual design takes place during the first two or three stages of this process and is usually conducted in a smaller time scale than the main definition phase. This allows many concept options to be examined and a knowledge database to be built up. This database may contain both technical and costing information such that a project may eventually go ahead with a high degree of confidence.

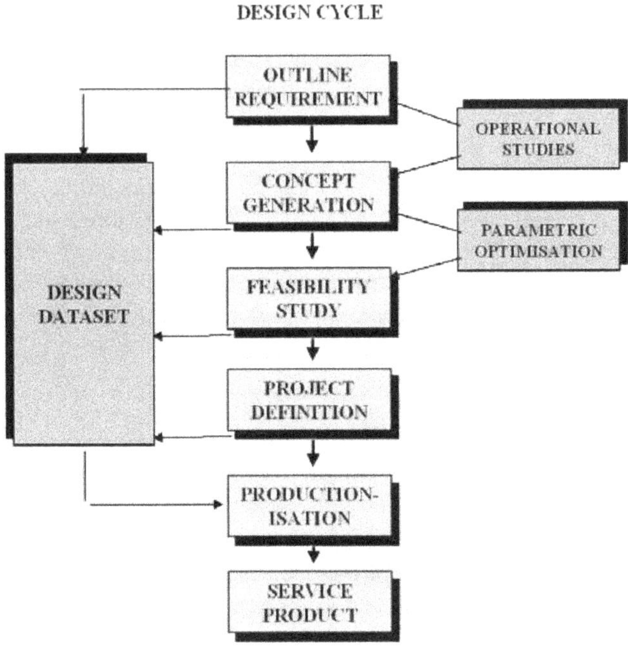

At the start of the process it is usual for only an outline of the requirement to exist. Requirements capture and functionality derivation results in the development of the specification, a process in which the customer may take part. Exact performance requirements, and subsequent design features, may only be generated later in the study and may involve much investigation and discussion.

The requirements capture process starts from a set of operational requirements, generated and substantiated through an operating scenario. The design of a complex product such as an aircraft, a ship or a car is usually dictated by a number of intended primary roles. Each of these roles provides constraints ('design cases'). The requirements are translated into a set of design configuration options which attempt to meet the requirements in various ways. Several iterations may then be required to select the final configuration. The final solution concept will include specific performance requirements, structure and systems required, and result in an outline shape, perhaps embodied in a parameterised digital model. Following this stage, more detailed design is required to provide a full engineering definition. Optimisation can play a role in the shaping of the detail design process itself. The internal layout of the product obviously depends on decisions about structural topology. A case study is included in section 7 in which a structural topology is optimised. Where this is carried out, it determines the internal volumes available. Therefore such an optimisation forms a design process stage in itself.

In each main phase numerous designs are investigated. The depth of each investigation is time bounded as well as dependent on the degree of automation available. The complexity of the total process is limited by the ingenuity of the engineers, the availability and accuracy of theoretical methods, and the cost and timescale constraints.

The requirements lead to objectives and constraints. The distinction between these is often rather hazy. Some constraints are precise and unvarying, such as those laid down by regulations. Others are softer and can be relaxed if necessary to satisfy other needs or to provide worthwhile gains. As requirements evolve and concept choices are made, some constraints will disappear and others will be introduced.

A classical problem at the concepts stage is to balance needs which drive down size, (e.g. minimise mass and cost) against others which drive up size (internal space needs for equipment, people, fuel, internal structure).

Once the solution concept has been sized and shaped a further level of detail refinement can be undertaken. This will include the layout of the major structural items and the installation of systems and equipment, and decisions about major engineering features such as engine installation.

3.3 Design As A Process

Design definition can be thought of in terms of successive transformations and additions of detail to the design data set, followed by 'freezing' of the design.

The natural 'freezing' stages are:

1. freezing the requirements
2. freezing the solution concept
3. freezing the product option specification
4. freezing the detailed engineering definition

Design 'freezing' corresponds to dealing with requirements in a hierarchical way. And to substructuring the problem by allowing different disciplines to work separately to a common baseline.

3.4 How Do People Design?

Evidently, during conceptual design, the design problem is complex and subject to change. It is not as amenable to a formal statement of the problem as later on.

WHY DO DESIGN OPTIMISATION?

Human ingenuity plays a large part in the search for solutions. It is clear that in broad terms design is a creative process involving the use of knowledge and experience. So the human designer is 'in the loop' of the search for an optimal design. Use of knowledge and experience avoids previous mistakes, leads to faster solutions and can lead to more innovative solutions. It enables solution of problems in which constraints and requirements are only partially defined at the start.

Current optimisation technology can tackle a useful and important, but restricted, set of problems. It is useful to know if the overall design problem is structured in such a way that the parts we can tackle are embedded into it. Are the parts which involve human selection different in kind and in the process used, from the selections which can be automated?. Do the selections which we can currently tackle automatically occur late in the cycle, i.e. are they of a more detailed 'low value' nature?. Are the broader design problem's difficulties such as to make automated search relatively small in value? What stages of the design cycle are the most relevant and amenable to multidisciplinary optimisation?

In recent years, there has been much research on how human designers work. One approach to provide insight on this topic has been to compare how novice designers work with how experts work, see for example [6].

This work supports the view that humans use a basic Generate-Realise-Evaluate process structure which is similar to that which underlies automated search. Experienced designers operate the realise/evaluate phases in a sophisticated and integrated manner, which shortcuts the need for full realisation and evaluation whenever possible, to home in on any reasons why a design can be eliminated quickly.

This suggests that solution methods can be structured so that the human searcher can be freely interchanged with automated search.

3.5 Optimisation Within The Design Cycle

We have seen that the overall design process takes place as a set of subproblems decoupled and idealised using engineering and project knowledge. At any stage of the process, the design subproblem being addressed can be optimised, subject to the following conditions.

1. The optimality criteria and the constraints are well defined and computable
2. Analysis codes to compute these are available and usable in batch mode
3. The machines on which the analysis codes are run can communicate

4. The design evaluation sequence is well defined

The case studies in section 7 are examples. They show that important high-value design decisions can be addressed and a major impact made.

4 Optimality Criteria

4.1 Optimality Or Design Improvement?

Although we speak of looking for an optimum design, in many instances our approach is slightly different. We start from an existing design, and our objective is to make a significant improvement to this. Usually this is because we are limited by time or money; or our knowledge of all the approximations which have gone into formulating the problem is such that we know a very accurate search is not justified.

The techniques available and appropriate for this type of search are not different in kind from what you would use for a full optimisation. But because of the different stopping criteria, and because you know you may well only be able to afford to look at a small number of designs, your selection approach is likely to be different.

4.2 Definition Of 'Best'.

Any criterion is adequate to define 'best', provided that it is consistent and computable.

1. The traditional measure of 'best' as the design with the greatest or least of a property of interest is the most straightforward.
2. 'Best' as greatest or least in a robust sense, i.e. insensitive to small changes in design variables, operation conditions, or other parameters, can also be used. Integrating over the range of the parameters in question, by using a weighted average, can be used. This is viable but probably considerably more expensive.
3. A pessimistic view of parameter values, e.g. materials properties, can also be taken, by adopting the 'greatest or least' criterion as in 1), but calculating the criterion with each critical parameter at its worst value, and taking the most pessimistic of the criterion values. This will also be more expensive.

4.3 Metrics Structure.

Optimisation requires you to decide on one or more metrics by which to judge your design.

Over the past few decades, it has been usual to optimise using a single metric as the objective, and many successful design decisions have been made by this means. An outstandingly successful area has been that of structural optimisation. Various metrics have been employed, chiefly structural weight. The case study of section 7.1 illustrates this.

If you are addressing a subproblem within the design, the metrics to use may suggest themselves. If you are trying to minimise drag, why not use drag as the metric?

However, most such optimisations have to be seen within the framework of the product as a whole. The customer for the product will see its qualities from an overall perspective. How much does it cost to buy and run? How well will it perform? Will it work in his environment on his job? How often will it go wrong? How much support will it need? How long will it last? Each of these questions can lead to a metric by which to judge the product.

High level metrics can be decomposed into other metrics, in a tree structure. The relationships are not necessarily simple and additive, but provided they are understood and computable they can be adopted for optimisation. An example tree, from the aircraft industry, is given in [7].

From a manufacturing point of view, process-based metrics can be seen in a similar way with regard to the cost of the process, the quality of the end product, and process time.

The adoption of a top-down metrics structure is strongly advocated as a means of focusing design optimisation on what are the real problems to be solved. Otherwise, you may spend your time trying to minimise drag from a technical perspective without a clear idea of whether from a business viewpoint you should be trying to minimise operating cost (i.e. miles per gallon) or maximise range (i.e. miles).

4.4 Tradeoff Decisions Using Multiple Criteria

In assessing a product, there is usually no single aspect by which it is judged. Cost, performance, operational suitability, durability, support; it is hard to add up these diverse factors into a single figure of merit. What happens in practice is that the designer balances these factors off against each other to arrive at what he thinks is the best combination of properties in the final design.

This balancing act is complicated. It can also of course be very subjective. Each factor is relevant in the sense that a very poor value may make the design unacceptable. But as far as the 'exchange rates' between improving one factor at the expense of another, things are not usually straightforward. One factor may be judged 'sufficiently good'; so there is no merit in sacrificing other things to improve it further. Conversely, there may be little value in a small improvement; but reaching a threshold value may suddenly give big benefits e.g. entry into a new market. So the process of using metrics in combination is complex and nonlinear.

The important issue here is to recognise that for many major problems, it is inadequate to fix on a single metric for design optimisation, or to combine metrics in a premature way. Means are needed to address problems with multiple objectives. Techniques for doing this are discussed later

.

5 The Anatomy Of A Design Optimisation

When you conduct an optimisation, you take as a basis a family of possible designs, and you consider as large a number of candidates from the family as you can, and try to pick the best one.

In doing this, you are performing three operations time and again. These are

1. step 1: selection of the candidate design, specified by a set of design variables, the design definition dataset
2. step 2: production of the detailed information needed for step 3
3. step 3: evaluation of the design in terms of the criteria being used to decide which is 'best'

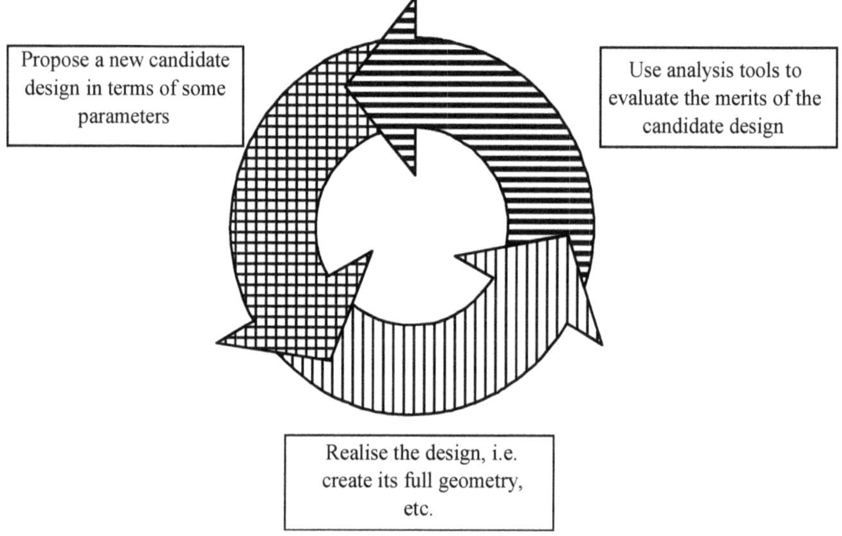

First, you select a candidate from the family. There exist nowadays many sophisticated search algorithms which can assist in doing this, or take charge of the whole search.

Secondly, candidate designs are usually characterised by a small number of parameters. For example, overall dimensions such as wing sweep angle, or fuselage length. So it is nearly always necessary to carry out an operation to "realise" the design by turning these few parameters into a complete definition, including geometry, materials, and other data.

This is because the solvers needed for the third type of operation, which is "evaluation" of the design, usually need this type of detailed geometrical input.

The context used for illustrations is engineering design, but the optimisation approach is applicable generally. The examples are mostly from product design, but process design in its widest sense is also treatable. To use it, you must be able to analyse. Many times in the past, and still in many contexts today, organisations do not include analysis as part of the design process. The process consists of design-test-modify, with the feedback from testing taking the place of analysis. Sometimes you have no option, sometimes you have.

5.1 Formulating The Optimisation Problem

For clarity and precision, we should state the optimisation problem class as follows.

Maximise $F_1(\tilde{X}), F_2(\tilde{X}), \ldots\ldots\ldots F_n(\tilde{X})$ with respect to the design vector $\tilde{X}$ subject to $c_i(\tilde{X}) \geq 0; \; i = 1, m;$ where $F_i(\tilde{X})$ are the objectives, and $c_i(\tilde{X})$ are the constraints.

In the generic optimisation process described above, the 'Propose candidate' step generates a value of $\tilde{X}$, and the 'Evaluate' step computes the values of $F_i(\tilde{X})$ and $c_i(\tilde{X})$.

5.2 Design Evaluation

5.2.1 Design Evaluation Sequence

Design evaluation is a cornerstone of optimisation. This is where engineering analysis tools are used to provide the quantitative metrics by which the design can be judged.

Because the results from one tool often provide the input to another tool, this evaluation activity needs to be considered as a process in itself.

Take the example below. The diagram illustrates a process used for aeroelastic optimisation by Daimler Chrysler Aerospace, of a wing flap which is characterised by three geometry parameters concerning flap angles and positioning.

It shows aerodynamic calculations which produce loads, which are passed to a structural computation which then sizes the structural members. And finally a calculation of the two metrics which are considered here, which are weight and roll rate. So this process can be used for any candidate design, to produce these two metrics.

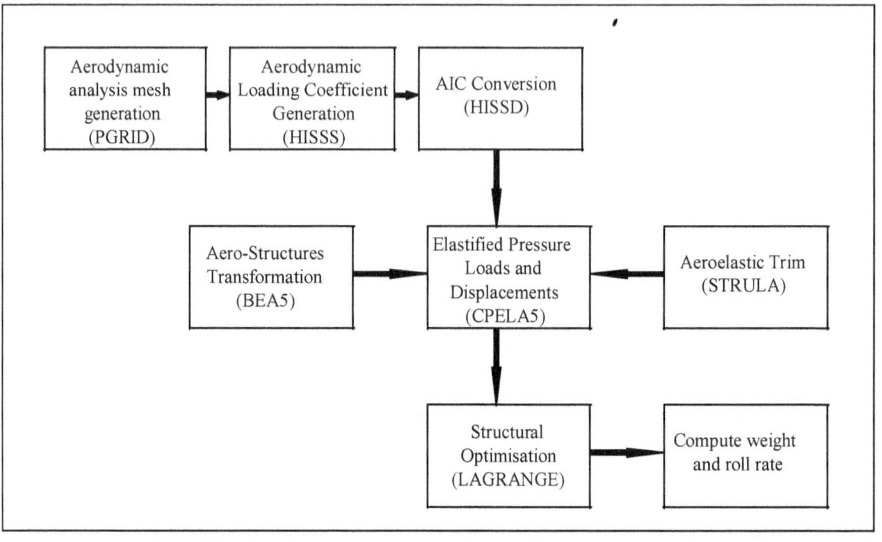

It is the design engineer's responsibility to define and provide this evaluation sequence, as well as the tools to carry out its individual stages. Within the sequence, all values of objectives and constraints have to be computed.

5.2.2 Design Evaluation Sequence Definition Requirements

To be able to compute a general design evaluation sequence in a fully automated manner, and to have the ability to freely define such a sequence at runtime, the following essentials must be provided to the designer:

1. ability to define which computing modules to run, in order to execute each individual process block in the block diagram
2. ability to define what platform each module is to run on
3. ability to define the sequence in which the blocks are used and what to do in case of failure

5.2.3 Application Code Validation

In an optimisation, the design search algorithm may elect to examine designs which are very diverse. This may mean that analysis application codes are asked to operate in regions which are unfamiliar and relatively untried. It is essential to consider in what ranges the application solvers being used are valid. In the course of an optimisation a large number of cases is analysed. It is extremely difficult to inspect each individual analysis to detect anomalous results. Therefore, a clear view is essential about what to do to intercept such cases and deal with them.

Otherwise there is a large risk either of breakdown of the optimisation, or of failure to detect misleading results.

5.3 Design Definition

The definition of the design to be optimised is quite likely to be held within a product data management system, and to have been produced using a standard CAD package.

Over the past few years the use of CAD has been introduced into the initial design process, originally in the form of automated 2-D draughting and subsequently with full 3-D modelling capability. The 3-D working capability of modern CAD systems has allowed designs to be produced with higher confidence than before. The designer can readily switch between 2-D and 3-D modes thus allowing the 3-D nature of the design problem to be fully explored. Outer surface continuity and smoothness can be better appreciated and confidence in the placement of internal components improved.

The use of a single model acting as a digital master also allows previously separate activities such as structural layout and systems installation to be combined. The individual designer is therefore responsible for all aspects of the initial layout and can use the same controlling geometry for each task.

The electronic storage of data now gives much more flexibility in the form of output that can be generated. Not only does it make interfacing with analysis packages much easier but increasingly potential customers are requiring design information to be supplied in tape or disc format in addition to the more traditional drawing set.

In considering options for implementing the automation which is essential for optimisation, we are likely to want to take one of two possible approaches.

5.3.1 Approach A: Analysis Models

The simplest approach is to work with a model of the design which is specially chosen to be suited to the analysis which is needed to produce the design metrics needed for the optimisation. For example, many research papers on wing design describe optimisation using a traditional wing section such as one of the NACA series. These are described by mathematical formulae. The shapes can be varied by varying the parameters in the formulae.

Analysis models have some important advantages:

1. parameters can be the most appropriate to the physics of the analysis

2. the use of such models is faster - there are no complex CAD updates
3. it may well be considerably easier to put the data into the form expected by the analysis tools
4. no idealisation is necessary

The disadvantage is that when the optimisation is finished, you have to take your design and input it into the CAD system, i.e. redraw it. This will be labour intensive and will result in an approximation which will need checking for accuracy and for preservation of essential relationships. Also, where there is more than one major analysis needed in the design evaluation sequence, it may be not be possible to adopt an analysis model which is specially suited to the physics of each of the analysis problems.

5.3.2 Approach B: CAD Models

If the design to be optimised exists initially as a baseline produced by a CAD system, and you want to ensure that any designs which the optimisation considers are fully compatible with the CAD system and capable of being stored and used by it, it may be possible to use the CAD model directly as the basis for optimisation. However, a number of issues arise.

Parametrisation

If you are to use the digital data model, it must be parameterised. At first sight this may not seem to be a problem, because CAD models are inherently parameterised. However, there are major difficulties in using basic CAD forms directly.

1. You need to understand the mathematical forms of the geometry entities being used; you need to be able to read the 'canonical form' data defining these entities directly; and you need to be able to change the data and reinput it into the CAD system
2. In carrying out such changes, you need to understand and be able to preserve the continuity conditions of the CAD model, under changes in the parameter values
3. Typically, there are large numbers of parameters in a CAD model, each affecting the design in a rather detailed way; instead of the much smaller number of more physically meaningful parameters which a designer may want to use in an optimisation
4. The CAD parametric form will not be optimal for modelling the physics of the analysis; this means that you will need to use a greater number of parameters, and therefore need more optimisation iterations

These difficulties are so large that in practice an end user of a CAD system who takes this route for providing a parametric model for optimisation would be most unlikely to succeed with the optimisation.

However, some CAD systems now provide for export and import of models in a parametric form, using parameters defined by the designer at the time the model is drawn. These models are much more suitable for optimisation, with some limitations expressed below. There may however be only a limited range of geometric entities which you can use as parameters. And any constraints which need to be preserved under changes of parameter values will have to have been foreseen and specified explicitly by the designer at the time the model is defined. Thus, converting existing models which have been drawn in traditional fashion into parameterised models is difficult.

Assembly modelling

There are many ways of using a CAD system to define a complex object. In practice, the product model is likely to consist of an assembly of several models of parts of the design. This is because it makes the design work easier to distribute and control. These models may number from one to several thousand. For optimisation purposes, we need to take the baseline product model, change the values of the parameters in our optimisation, and export from the CAD system the data required by the analysis tools. If our design is actually an assembly of a number of models, we shall need the parametric capability of the CAD system to operate on the entire assembly. This may or may not be possible within the capability provided.

Data export/import

To operate the process of making the changes to a CAD model which are needed to reflect a change in parameter values, you need to be able to carry out two types of information exchange between the optimisation system and the CAD system:

(1) Import the new parameter values into the CAD system, and use them to update the CAD model.

(2) Export from the CAD system the data needed by your analysis code to describe the new design, in whatever data exchange format (IGES, native form, STEP, other) it uses; systems may vary in the access methods they provide for this (files, database access, or objects).

Data exchange of a non trivial design is likely to provide some significant performance issues. Amounts of data which may run to several gigabytes could be involved. So this should be carefully considered before committing to the use of this approach.

5.4 Idealisation

By idealisation we mean simplifying features of a design by eliminating elements which do not affect the properties of interest, or substituting a simplified model.

Why is the topic of idealisation relevant to optimisation? Because it is rarely automated, and is often very labour intensive and very time consuming. There are various examples where the idealisation of a design in order to render it suitable for analysis has taken months or even years to carry out.

Some types of idealisation for analysis purposes tend to be associated with qualification checks on a design which is at a late stage, i.e. it is relatively mature and possesses a considerable amount of detail, which needs eliminating if the check analyses are to be reduced to practical proportions.

However, other idealisations of major importance for optimisation fall within the 'Realise' block of the trio of optimisation actions described above. These include:

1. selection of surfaces which define the outside of a vehicle, preparatory to fluid dynamics analysis
2. repair of geometric surfaces to ensure there are no gaps which would compromise electromagnetic analysis
3. preparation of a structural finite element model from the structural design

To promote optimisation, such activities need to be automated. There may well be ways of utilising CAD capabilities in ways different from usual, to facilitate this; for example by adopting design practices which creating outside surfaces as clearly identified entities which can be accessed directly, instead of as the result of a laborious process of intersecting surfaces and removing internal items.

5.5 Systems Integration.

5.5.1 Industrial High Level Requirements.

For a number of reasons, industrial design organisations have acquired a strong motivation towards integrated design activities in recent years.

With respect to computing support for design:

1. there is now an expectation of achieving software solutions which have a high degree of computing platform independence, leaving the user organisation free to buy in the most cost effective hardware solution of the day.
2. the extremely high cost of employing inhouse software development teams to produce bespoke solutions has resulted in a move towards use of off-the-shelf analysis packages wherever possible. In the past, when a

need for interfacing codes arose, usually just two codes were involved, supporting some particular small part of the design process. The codes were more likely to be inhouse codes, and there was always a tendency to approach the problem by changing the codes. These days, employing 'black box' bought-in codes, and with more codes likely to be involved, it is apparent that a more general interfacing and integration approach has to be adopted.

3. in most engineering disciplines there are several alternative solvers to choose from. So it may be desirable to plug-and-play with different options.
4. there is a need for analysis codes to be able to operate together regardless of where they are on the network.

Other reasons arise from current trends and imperatives of industrial competition and collaboration.

1. Many design organisations are trying to move towards an 'Integrated Product Design' process. Analysis needs to be integrated into this process.
2. As product competition increases, there has to be a move in the direction of focusing on the product's impact with the customer. Thus the product needs to be constantly evaluated during its design in an overall sense, measuring it in the same ways the customer will, rather than concentrating on lower level technical properties.
3. As well as industrial competition, we are nowadays seeing many global collaborative projects taking place, with design teams dispersed in several countries.

There are thus many motives nowadays for wanting a general way of integrating the operation of design applications.

5.5.2 I.T. Requirements.

So what are the implications of these industrial requirements in terms of I.T?

The platform independence issues are fairly clear. A 'platform' is a combination of the hardware and the operating system. Whether a particular application will run on a particular platform is a matter for the application's producer. The main platform issue for optimisation is whether applications on different platforms can communicate with each other.

The I.T. issue here is therefore *what technologies to employ* to directly address the need for applications to communicate over a network.
The use of 'Off the shelf' components within design optimisation implies the following technical capabilities:

- *runtime definition of component identities*; in principle, the user should be able to change any computational component in the design evaluation at will
- *runtime definition of platforms and paths*; the user needs to be able to say where on the network each computational module is, and how to activate it
- *runtime definition of the design evaluation sequence;* the computational modules will need to operate in some cases in a particular order, so the operating sequence must be specified: defining the sequence requires
    - an input medium to use for the definition (graphical, text, other)
    - a 'language' in which to describe the events in the sequence
    - an event-based activity handler, to take charge of the running of the sequence, and the handling of any exceptions

5.5.3 Technology Options For Meeting The Requirements

Many engineering enterprises use applications running on a variety of platforms. Thus some generality is desirable in the integration methods used. We therefore choose not to discuss approaches which are manufacturer-specific, such as Microsoft tools like ActiveX, DDE, COM and OLE.

The issues to be addressed are mainly

1. network communication
2. distributed application programming
3. interfaces between existing applications
4. object management

At the networking communication level, connectivity between most platforms has been available for some years. This has not been widely exploited until recently, due to lack of standard interfaces between applications.

Two technologies which are nowadays available on any significant platform, and which provide a solution to wide area inter platform communication are Java and CORBA.

Java was conceived as an object-oriented programming language. Inherent provision is made for the programming of applications distributed across a network through features such as URL and Socket classes, which make it easy to download an object or open streams to read to and write from it. A Remote Method Invocation API allows a JAVA program to invoke methods of remote Java objects as if they were local. More recent features have widened the capabilities to manage

Java objects, through the provision of naming, directory and security facilities. Java has inherent portability through its use of byte-codes interpreted through the Java Virtual Machine rather than native machine code. This facilitates systems integration through avoiding implementation problems and delay when new platforms need to be used. Java is referenced in [8].

CORBA is an industry standard which defines an architecture and facilities for distributed object-based computing. Interoperability of diverse applications is handled through interfaces defined in Interface Definition Language (IDL). Legacy applications need to be wrapped in order to provide such interfaces. Objects are used across the network through the agency of an object request broker, which takes care of locating objects and activating servers. A comprehensive set of object services and common facilities has been defined using the interface standard. These provide for naming, event handling, security, licensing, and database storage amongst others. Commercial implementations of CORBA differ somewhat in the range of facilities offered. CORBA is referenced in [9].

5.5.4 The Role Of Parallel Computing

In the quest for greater computing power, most large machines nowadays feature parallel processing in some form. Parallel computing can play a major role in optimisation, in two ways. One is the traditional way, in which a 'parallel version' of an analysis algorithm is used to provide a solution in a faster clock time, at the expense of more overall processor power. The other way uses the natural parallelism inherent in some optimisation algorithms. Earlier, we described the 'propose-realise-evaluate' sequence which is carried out on any design considered during an optimisation. When a number of designs are being considered simultaneously, they can be evaluated in parallel. This is particularly appropriate, for example, when using a genetic algorithm search, described later. For major engineering optimisations, the availability and exploitation of parallel computing capabilities is indispensable for practical work.

5.6 Defining The Family Of Designs To Be Searched: Parametrisation

In paragraph 5.1 we defined what we mean by an optimisation problem. In this definition, we are concerned with searching through a family of candidate designs. The variables which characterise the design are contained in the parameter vector $\tilde{X}$. The individual parameters can be either continuous variables, such as geometric dimensions, or discrete variables, such as the number of an item in a catalogue of off-the-shelf components. Problems can of course also have a mixture of continuous and discrete variables.

5.6.1 'Continuous Variable' Problems

If there are n continuous variable parameters which together define an individual design, the space containing all possible designs in the family is n-dimensional, and each design can be thought of as one point in the space. We did not previously use a notation to denote the realisation of the design itself; we will call this $D(\tilde{X})$. What we denote by this will usually be the design's full geometry, but it could include other aspects as well or instead. It is important to understand that $D(\tilde{X})$ is only a model of the 'real world' design space which the designer would like to be dealing with, and it will have limitations in what it can describe. For example, if we want to design a simple 2-D closed boundary, we could take as our parameters the variables $\tilde{X} \equiv (a,b,r)$ and the boundary as all points (x,y) on a circle centred at (a,b) with radius r, i.e. $D(\tilde{X}) \equiv \left\{ x,y \mid (x-a)**2+(y-b)**2=r**2 \right\}$. However, we would need to understand that there is no way in which our boundary could ever represent some of the shapes we might want to consider; for example it could never have corners.

In choosing the model form, the user has several factors to take into account.

5.6.1.1 Suitability For Modelling The Physics

A main issue for design is choice of a model form which is capable of giving good design performance. An obvious example case is aerofoil design. Decades of work have been done to find families of aerofoils which can provide required lift performance whilst minimising drag. Even a small amount of drag saving will make a major impact on aircraft operating costs. So we might for example choose to adopt a NACA thickness shape for the aerofoil, given by

$$y=5t*(0.2969\sqrt{x} - 0.126\,x - 0.3516\,x**2 + 0.2843\,x**3 - 0.1015\,x**4).$$

This model has one parameter, the 'maximum thickness'' t. Clearly, there are many thickness distributions which this family can never represent. So, depending how sensitive the performance is, the choice of model form may be crucial. It is important also to consider that a model form which is well suited to one design discipline, e.g. aerodynamics, may not be at all suitable for producing an acceptable design for other disciplines, e.g. radar returns. Thus, model choice becomes more of an issue as multidisciplinary design practices are adopted more widely.

5.6.1.2 Suitability For Optimisation

Adopting a model with a large number of parameters may appear to give more freedom of choice of the final design. However, the more dimensions the parameter space has, the more work there will be in searching the space for optimum designs. In practice the work, and therefore the computational cost, snowballs as the number of parameters increases. Since the computation is quite likely to have to be guillotined for cost reasons, the choice of a model with a large number of parameters may actually result in a very superficial search of a large design space, leading to a less effective result than if a well-chosen model form with far fewer degrees of freedom had been adopted.

5.6.1.3 Suitability For Feeding Back Into The CAD System

If the user opts to use the CAD model directly, as discussed in section 5.3.2, a further issue is present. CAD systems offer a range of geometric forms for the designer to use in constructing models. The usual 2-D and 3-D forms available consist of piecewise spline or polynomial functions. For the reasons discussed in 5.3.2 these will need to be specially parameterised for use in optimisation. Also, the user will need to consider whether CAD model forms are suitable for the physics of performance analysis as indicated in 5.6.1.1.

5.6.2 'Discrete Variable' Problems

Many times when we make a choice, we have a list of individual options to pick from. In selecting a material, the options might be steel, aluminium or plastic. We could identify these as options 1, 2 and 3. There are no 'in betweens'. In such an optimisation problem, the variable representing material would be discrete rather than continuous. This situation happens also when we are putting together a system design which has to use off-the-shelf items selected from a catalogue. If we are designing a hydraulic system, say, we are likely to choose a pump from those already on the market, rather than design one specially. In this case, the pump options could be numbered discretely.

There are no parametrisation problems with discrete variables. Discrete variable problems are easy to set up. Their difficulties lie in the search process. There may well be no meaningful concept of 'nearness' of options to each other. The performance of a steel item which we happen to have numbered 1 will not be closer to an aluminium one numbered 2 than to a plastic one numbered 3. So we cannot search using concepts of good regions and good directions.

5.6.3 'Mixed Variable' Problems

In practice it is quite common for problems to have a mixture of continuous and discrete variables. This does not pose any extra difficulties in setting the optimisation problem up. It may however restrict the user's choice of search algorithms somewhat.

5.7 Design Search.

Design search is a process of intelligent exploration of the objective and constraint spaces. By using the results of evaluating those designs we have already considered, we aim to forecast where better designs are to be found. To quote one eminent engineer: "Success can be achieved only through repeated failure and introspection." (Soichiro Honda).

Search methods have been developed following two widely different approaches.

Firstly, many deterministic methods have been proposed [10] [16], usually based on assuming that the objective function being investigated is continuous and reasonably smooth. These methods are often called 'hill climbers', because in seeking to maximise the objective, they try to move 'uphill'. Some of these methods estimate or calculate slopes and curvatures; some use the idea of quadratic approximation near the optimum; some use 'direct' logical search moves based on previous evaluations.

In deterministic methods, the searcher starts from an initial design and makes a sequence of moves from one design space point to another, to try to improve. The process is terminated when there is evidence that the maximum has been reached.

In the second family of methods, the searcher uses a set of initial designs. Then, using methods which are based on analogy with mechanisms of natural evolution, new generations are created successively through a 'breeding' process. At each stage, the best individual design can be identified and accepted when the searcher decides to stop the search. The methods use random selection of 'parents' employing probabilities which balance the retention of previous 'good design' features against the selection of radically new designs. The most widely used of these methods are 'Genetic Algorithms' [11].

5.7.1 Finding A Feasible Solution

In some problems the constraints which the solution has to satisfy are very restrictive. It may be far from obvious whether the problem has a solution at all. This is particularly possible in problems with large numbers of constraints and variables.

It is obviously useful to know if there is no possible solution to the problem as formulated. It is also useful to be able to play around with the constraints to find out which are the most restrictive, and decide how to weaken them to make the problem solvable.

For most optimisation problems it is useful in any case to have a starting design which is feasible. And for some problems, for example scheduling problems, it may be sufficient to find any solution which is feasible.

Constraint programming [12] is an emerging technology for solving such problems, especially applicable in areas like scheduling and circuit design.

5.7.2 Preliminary Exploration

Before employing a search strategy such as described below, it may be useful or essential to carry out a preliminary exploration of the design space. This might be in order to provide an initial 'population' of candidate designs. Or to let the user build some understanding of the behaviour of the objectives and constraints, prior to deciding what further search method to use.

A range of reasonable ways exists for positioning a set of N designs. These include

user-chosen set:	based on the user's previous experience
random choice:	eliminates subjective bias, at the expense of higher risk of uneven sampling
number-theoretic methods:	these originated in the problem of numerical integration of a function, where the function's values at a set of 'representative' points are required; various methods exist, e.g. Sobol' algorithms [13]
planes of experiments:	various systematic ways of positioning candidates are available; e.g. grid-based methods, suitable for low number of design variables (cell centres, face centres, nodal positions) : Design of Experiments methods [14]: Taguchi methods [15]

5.7.3 Single Objective Maximisation

Problems where there is only one objective to be optimised can immediately be subdivided further. The 'Continuous variable' problems described in section 5.6.1 are either Unconstrained or Constrained.

Selection of a search method for unconstrained problems depends on the nature of the objective function [10]. Chief factors are:

1. Is the objective a continuous smooth function?
2. If not, it is usual to adopt a 'direct search' method, i.e. one which does not make any assumptions about the function or attempt to estimate its gradient and use it to proceed uphill.

3. Is its behaviour smooth and unremarkable enough to assume quadratic behaviour in the region of the optimum? If so, a conjugate direction search could be appropriate.
4. Can we compute its gradient? Can we compute its second derivative? We can choose from a variety of gradient-based techniques which use the 'steepest descents' direction or the 'Newton' direction or a combination, either by computing the first and second derivatives directly, or by approximating them.

When constraints are present, they can be dealt with in various ways. Some methods try to ensure that all designs considered satisfy the constraints, and search by moving around the 'constraint boundaries'. Others attach 'penalties' to designs depending on by how much they violate the constraints. Deterministic methods are surveyed in [16].

The advantage of deterministic methods is efficient local convergence, when the search is near the optimum.

When the objective function has properties such as noise, discontinuities or many peaks and valleys, it is much more likely that probabilistic methods, which are expensive but much more robust, will be appropriate. A tutorial on basic genetic algorithms is provided in [11].

Genetic algorithms can be used to treat both continuous and discrete variable problems. Discrete variable problems can also be addressed through various methods such as Cutting Plane or Branch-and-Bound, known collectively as 'Integer programming' methods. Description of such methods is given in [17].

5.7.4 Advanced Search Methods

For many engineering design problems, today the computational cost of one design evaluation is so large that the number of evaluations which can be afforded is very limited. Standard search methods explore the design space using methods which are based on assumptions about objective functions' behaviour. However, when evaluations are expensive there are strong reasons for augmenting the standard search methods with any other techniques and sources of knowledge available, to better describe the space being explored.

5.7.4.1 Response Surface Methods

Traditional search methods referred to in 5.7.3 often assume an underlying mathematical form of the objective; almost always assuming quadratic behaviour. At each step, they use latest evaluation information to deduce an approximating quadratic, then use the maximum of the quadratic, which can be computed immediately, as a next design point for evaluation. For 'expensive to compute' objective (or constraint) functions, it is worthwhile making much more effort to model current evaluation information, and then carry out a search on the relatively 'cheap to compute' model to determine one or a set of next design points for evaluation. The model is known as a 'response surface'. All the traditional forms of mathematical interpolant can be used. More recent nonlinear forms have also been adopted. Neural net approximants have proved effective [21]. The statistically based 'Gaussian processes' method can also be extremely productive [22].

5.7.4.2 Rule Based Methods

In complex engineering problems, when time is short and not many design evaluations can be undertaken, there is every incentive to make use of the engineer's knowledge and experience to supplement or guide the search. In addition to constraint rules which exclude parts of the design space, rules can be envisaged which restrict the search to a small number (or one) of the variables at the current point, because these are known to be the most effective (or only) influences [23] or rules on which objectives to improve can be supplied [23],[18]. Rule based methods can be combined effectively with other methods to take advantage of the strengths of the gradient-based and probabilistic approaches [23].

5.7.4.3 Multilevel Methods

Multilevel methods have been introduced in recent years. They have some similarities to response surface methods, in that the search process switches between different design evaluators from time to time; and have a similar motivation to make best use of costly design evaluations by providing an alternative cheap evaluator. In this case, the cheap method is a different solution method which happens to be computationally simpler. In this approach, the simple method thus adds information in its own right.

5.7.5 Multiple Objective Maximisation

The multiple objective maximisation problem has been stated in section 5.1.

An 'ideal design' would be one which gives the same objective function values as would be obtained if each objective were to be maximised on its own. Such a design is unlikely to exist. Rather, there will exist a set of 'best designs' at the boundary of the feasible region. This boundary is termed the Pareto boundary, or Pareto set. Each design in the set corresponds to a particular set of relative weights associated with the objectives. Moving from one design to another corresponds to changing the relative importance of the objectives, in other words trading one off against another.

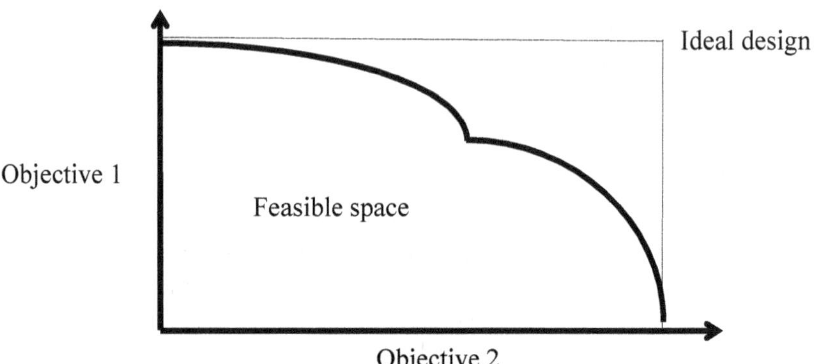

In practice we want to end up with one design rather than many. So the process of clarifying the decision maker's (DM's) preferences is a central feature of multiple objective maximisation methods. This fact provides a way of categorising these methods in terms of when tradeoff information is received from the DM in relation to when the set of optimum solutions is generated. The categories are

1. methods which articulate preference information before generating solutions ('A Priori' methods)
2. methods which articulate preference information progressively during generating solutions ('Interactive' methods)
3. methods which articulate preference information after generating solutions ('A Posteriori' methods)
4. methods which do not articulate preference information at all.

Each method is usually a combination of

1. use of preference information to formulate a problem with a single 'figure of merit' objective function which can be a linear combination of objectives, or a utility function [18]
2. a search method

'A Priori' methods

In these, the DM states some preferences before searching for a solution or a set of solutions. He may for example prescribe a fixed set of weights w_i to attach to each objective, then search for a minimum of the composite objective function $\sum_i w_i \; F_i(\tilde{X})$. Or he may define a priority ordering of the objectives, and maximise according to this (Goal Programming [19]).

'Interactive' methods
In these, the DM's preference information is used to guide the search. For example to define a cone corresponding to an interval of weight vectors, by sampling, then shrink its size progressively (Interactive Chebychev method, Interactive Weighted Sums method [19]).

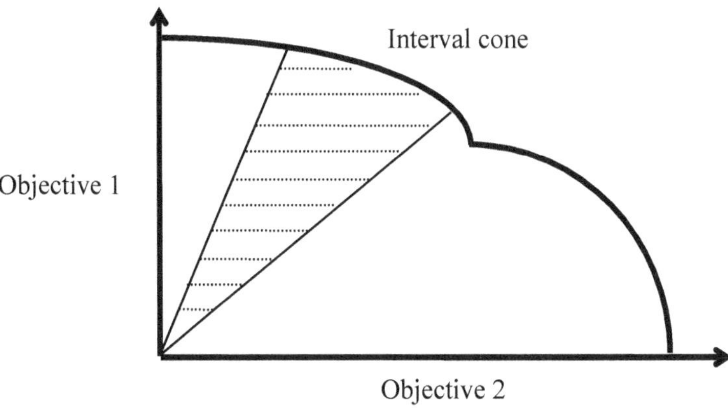

Or by setting up a point on the Pareto boundary, by putting lower bound constraints on all objectives except one, which is then optimised (e-Constraint method [19]). Then getting the DM to provide weights to determine which tradeoff direction to move in, giving new values for the constraints (Interactive Surrogate Tradeoff Method [20])

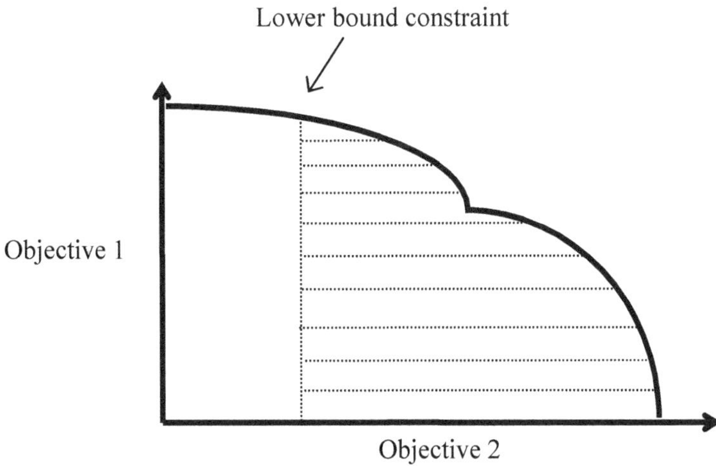

Or by using the DM to define target reference points in his area of preference and projecting them onto the Pareto surface.

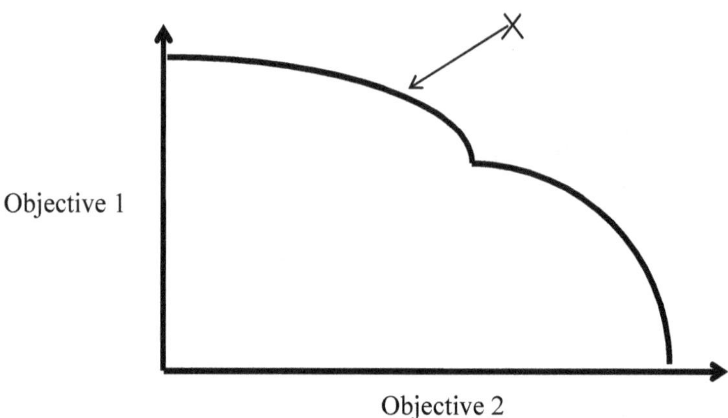

Or by using the DM inside the search loop of a non linear optimiser, and employing him to provide search direction information based on his preferences, in place of gradient values (Geffrion-Dyer-Feinberg method [19]).

'A Posteriori' methods

These include any methods which generate a set of solutions, with the DM selecting a preferred solution afterwards. Examples are the e-Constraint method used with a systematic variation of the constraints, and the weight vector method used with a systematic variation of the weights [19].

'No preference' methods

If the problem is formulated to use a simple predetermined combination of the objectives, for example the 'method of global criterion' [20], no user preference information is involved.

Search methods

Many of the available methods were originally proposed for linear objectives and constraints, and were based on searching via linear programming, but can also incorporate almost any traditional nonlinear (single objective) optimisation method. Some methods depend on using gradient information, which is unlikely to be readily available in most industrial problems.

Multiple objective genetic algorithms can be used for the search. These will by definition be an 'A Posteriori' method. However, they can be regarded as a means of generating information for the user to base preference information on. Their use can then be regarded as a preparatory step, before going on to use an 'A Priori' method.

References [18], [19] and [20] can be consulted for further information.

5.8 Decision Support Aids.

5.8.1 Visualising Analysis Results For Verification And Computational Steering.

During design evaluation, it is very likely that the design team will want to check the analysis results produced. For large complex field calculations, for example fluid flows around complex bodies, there is a need to assess and appreciate what is happening over the whole problem, to verify that there are no anomalous features of the calculation. Such verification may well be enormously complicated and require the highest human expertise. Thus, the use of 'Immersive visualisation', or 'Synthetic Environments', is presently of growing importance. As well as results verification, such expertise can also be directed to steering the optimisation calculation by revising the optimisation problem parameters, or by directly choosing designs to be assessed.

5.8.2 Visualising And Understanding Tradeoffs.

5.8.2.1 Picking Out The Pareto Set

At any given point during an optimisation, a set of feasible designs will have been evaluated. It may well not be known whether any of the designs is on or near the actual boundary of the feasible region. If the number is at all large, it will not even be obvious which of the designs are Pareto-optimal with respect to this particular set. So a filter needs to be provided to identify the nondominated members of the set. These can then be listed numerically or displayed graphically.

5.8.2.2 Solutions Which Do Not Satisfy The Constraints

In a problem with constraints, and depending on the optimisation method used, many of the designs evaluated may not satisfy the constraints precisely. This is particularly likely where equality constraints are concerned; for example, in an aerodynamics problem aiming to minimise drag subject to an equality constraint on lift. It is not useful to disregard such cases, because doing so may leave few or no others to consider. To benefit from understanding the results, the constraint sizes need to be displayed, or the results projected onto the constraint surfaces.

5.8.2.3 Picking Solutions From The Pareto Set

Various methods have been used to display sets of solutions in a multiple dimensional objective space. When there are many objectives, a main diagrammatic tool to assist understanding is parallel coordinates [24], also termed value paths [19], with petal diagrams [18] as a variation. An example such plot is given in Case Study 4.

5.8.2.4 Relating Results In Objective Space And Design Space

The designer needs to be able to understand how the design itself is changing as he moves from one alternative in the Pareto set to another. It is theoretically true that points which are close to each other in the objective space plot need not be in any way related in design space. However, there is a predisposition to believe that in a problem where the design variables are continuous, and the mapping to objective space is continuous, the process of moving around the Pareto boundary in objective space will correspond to a continuous movement in design space. Thus the relationship of objectives and design variables around the boundary is significant in understanding tradeoffs. In many objective dimensions there is no single plot which can easily depict this. However, in the important case where there are only two objectives being traded off, a plot of objective values against design variables as the boundary is traversed is valuable. Such information can allow confidence to be established that the result is understood and robust. And it can also assist in steering the optimisation by enabling the designer to nominate other design solutions for evaluation, to fill in the gaps between solutions already evaluated.

5.8.3 Visualisation For Results Compression And Storage

It is not easy to appreciate just how large are the amounts of results data which a design optimisation can readily generate. A design evaluation which involves a time dependent fluid dynamics calculation with a very modest grid size of a few thousand elements can generate several gigabytes of results for just one design evaluation. If you need to evaluate a substantial number of designs, you may well wish to preserve the results for future reference rather than throw them away. The total storage space required for this is highly unlikely to be available, even for a short period. So the preservation of results by means of visualisation is an important possible option which should be considered from the outset of the optimisation task.

5.9 Issues In Practical Problem Solving

Many particular issues arise in the setting up and running of an optimisation problem. Some types of problem can be expected to arise in many or most problems, and it is therefore worthwhile having an awareness of these.

Parallel running
Design evaluations often involve the use of commercial codes. Performing evaluations in parallel may well involve either living within the limits of a (small) finite number of application licenses or investing in a greater number of licenses.

Application exception handling
In order to fully explore design possibilities, search algorithms tend to propose unusual designs, which strain or exceed the boundaries of applicability of the codes used for design evaluation. So, it is essential that these application codes should detect when their results are invalid and return error information; and that the optimisation user should arrange to trap such errors and return meaningful information. Operational failures must also be trapped; for example the case where no license is available. Otherwise, an optimisation may have very little chance of running through to completion.

Restarts
A major optimisation may well need many days of elapsed time to complete. 'Checkpoint and restart' capabilities in some form are essential in this case. Operational procedures and 'shift lengths' are also important.

Parallel evaluations using a single file system
When running several instances of an application code and its script in parallel, possible directory and file name duplication must be considered. If the machine has a single file system, scripts must be generalised, or copied to a working directory for each instance.

File writing delays
When one application is supplying results files as input to another in the evaluation sequence, it is vital that the files have been written and closed before they need to be read. Asynchronous operations may mean delays. It is essential to check file existence explicitly.

Results files
Optimisation problems will typically involve using commercial analysis codes, whose output is provided as results files. Quite apart from the possible complexities of handling file formats which are problem dependent, an accuracy limitation may well arise if the results are provided to a limited number of figures of accuracy. Comparisons and differencing which occur within the optimisation will be coarser and less reliable as a result.

6 Industrial Process Benefits And Issues

6.1 Optimisation Promotes Automation

It is generally recognised that the direct use of the main design data model as a unifying agent in the design process forces it to become more seamless and integrated. This in turn tends to force clarification of the process itself, and a reduction in cycle time. In order to carry out any significant design optimisation work, an automated design evaluation capability is a prerequisite. A move towards optimisation therefore reinforces this beneficial trend.

It may be thought that the direct use of the design data model for optimisation is problematic, because currently CAD data transfer is rather slow. However, in most optimisations the amount of data being moved may well not be so large as to dominate the calculation, compared with evaluation. There is also one aspect which is an advantage. In traditional CAD work, many of the problems associated with CAD data transfer arise through transferring from one CAD system to a different one. But in the data transfer needed in optimisation we are transferring from the CAD system to itself. So many of the problems and fixes, such as having to remove CAD elements which have been found to give compatibility problems, are not relevant in optimisation.

6.2 Optimisation Promotes Focus On Metrics

In section 4.3 it was argued that the clarification of a design's measures of success is essential to ensuring that the design team are addressing the right set of problems for the business. The use of optimisation ensures that this issue is brought clearly into focus, and that the sometimes very difficult task of ensuring that sensitive areas of the problem receive their rightful attention.

6.3 Optimisation Promotes Closing Gaps In Design Evaluation Toolset

The use of top-down design metrics implies a completeness of capability to evaluate these metrics. Not all areas of design are currently free from difficulties in this respect.

Analysis challenges: many engineering analysis problems exist, for example in fluid dynamics and electromagnetics, for which only a limited solution capability exists due to their needs for amounts of computing power and resource which are not available for everyday needs

Modelling challenges: various areas of design exist in which there are still significant challenges in developing clear agreed models of the design attributes concerned; example areas are costing and systems modelling

Data challenges: the complete data model of a major engineering product such as an aircraft or a ship is currently much too large to be convenient for participating in 'whole design' computations

The use of optimisation enforces a need to provide design evaluation tools in all the areas of design which are key to the important design choices. The requirement is not to make large leaps in technology overnight, but to fairly capture the ways in which the design is evaluated in practice by the design team and formally acknowledge these.

6.4 Optimisation Promotes Design For Analysis

To automate idealisation, and provide closer integration between 'The Design Office' and the analysis departments (Aerodynamics, Structures, Electromagnetics, etc), the activities of

- definition of the outside of the product
- geometry repair
- finite element model preparation
- elimination of detail

have to be closely considered as part of the design process, rather than as unscheduled labour intensive and time consuming procedures. It is typical for design data to be created to a standard which is unfit for any of the above operations. Without automation of these, full use of CAD models cannot be achieved, and even the use of analysis models will be less effective. The use of optimisation will undoubtedly bring these automation issues into focus.

6.5 Optimisation Promotes Definition And Use Of Operating Scenarios

Top level design metrics must inevitably reflect how the product is manufactured, operated in the field, and supported. The production and operating scenarios provide a framework to bring into focus the important cost, performance and customer satisfaction issues. The scenarios to be defined may need to encompass any or all of the following areas.

1. strategic studies affecting procurement
2. operational analysis of in-service use
3. manufacturing cost models
4. business models: (single company site; dispersed company; collaborative project)

6.6 Optimisation Provides A Means Of Rapid Organisational Learning

Often, an organisation's know how is what keeps it in business. This know how is the collective distillation of experience from having done similar jobs before and built up knowledge of the designs. Design optimisation, which allows thousands of designs to be experimented with on a computer, speeds up the learning process. By trying out designs which are diverse, it promotes a broadening of knowledge of the solution possibilities and increases engineering understanding. If the results of investigating a broader design range are condensed and stored suitably, an advanced knowledge base can be built for future reference.

6.7 Optimisation Provides A Means Of Outperforming The Competition

For many industries, time to market is a key issue. The ability to consider and evaluate many designs in a short period of time may confer a decisive advantage over the competition. See the Team New Zealand example quoted earlier.

7 Case Studies

The case study summaries which follow give an indication of the kind of design problems which are being addressed using optimisation today. Detail is necessarily limited. More comprehensive accounts of these cases can be found in references [25] to [31].

7.1 Case Study 1: Three Bar Truss Design

A three bar truss is shown below. The bars are of constant cross section, and the aim is to determine the cross sectional areas which will result in a minimum weight structure, subject to various constraints.

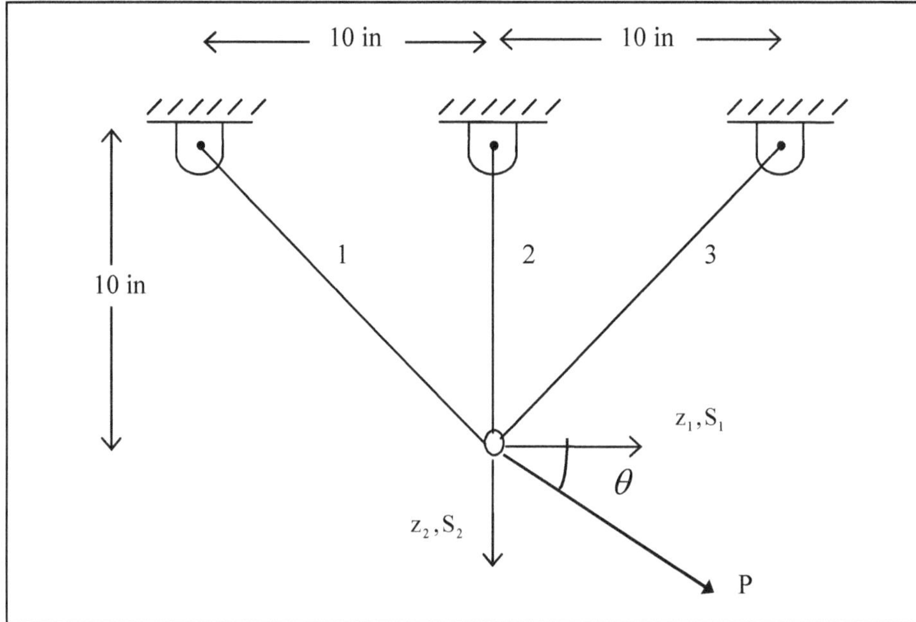

In general, constraints may be required on any of stress, buckling, displacement or natural frequency. The case considered here uses only stress constraints, which are formulated below.

Stress $\underset{\sim}{\sigma} \le \underset{\sim}{\sigma}^{a}$ where

$$\underset{\sim}{\sigma} = \begin{bmatrix} \sigma_1 \\ \sigma_2 \\ \sigma_3 \end{bmatrix} = \begin{bmatrix} E(z_1 + z_2)/20 \\ Ez_2/10 \\ E(z_2 - z_1)/20 \end{bmatrix}; \qquad \underset{\sim}{\sigma}^{a} = \begin{bmatrix} \sigma_1^{a} \\ \sigma_2^{a} \\ \sigma_3^{a} \end{bmatrix}$$

and the strains $\underset{\sim}{z} = \begin{bmatrix} z_1 \\ z_2 \end{bmatrix}$ are given by $\underset{\sim}{K}\underset{\sim}{z} = \underset{\sim}{S}$ where

$$\underset{\sim}{K} = \left(\frac{\sqrt{2}E}{40}\right)\begin{bmatrix} (b_1 + b_3) & (b_1 - b_3) \\ (b_1 - b_3) & (b_1 + b_3 + 2\sqrt{2}b_2) \end{bmatrix}; \qquad \underset{\sim}{S} = \begin{bmatrix} P.\cos\theta \\ P.\sin\theta \end{bmatrix}$$

Summary of case structure:

The structure has to deal with three loading cases; (P_1=40000, $\theta_1 = 45^0$) : (P_2=30000, $\theta_2 = 90^0$) : (P_3=20000, $\theta_3 = 135^0$) . This leads to 9 constraints. The constant values used are as follows. Young's modulus E = 10^7 p.s.i. Stress limits σ_1^{a}=5000 p.s.i., σ_2^{a}=20000 p.s.i., σ_3^{a}=5000 p.s.i. Density ρ = 0.1 lb/in^2.

Objective: mass of structure $= \rho(10\sqrt{2}b_1 + 10b_2 + 10\sqrt{2}b_3)$

Design variables: cross sectional areas b_1, b_2, b_3 of the members

Design evaluator all evaluations of objective and constraints can be directly coded

Optimisation approach

The problem has a linear objective function. Within the design space of all possible points (b_1, b_2, b_3), the feasible region formed by the stress constraints is interesting and presents a challenging search problem. There is a convex main region A of feasible designs, all corresponding to trios of non zero cross sections. If the cross sections are allowed to be zero, i.e. if 2 bar designs are allowed, then because the three constraint equations associated with the removed bar disappear, a new set of feasible designs exists corresponding to the regions B,C or D depending on which bar is removed. The region A approaches these other regions asymptotically at large cross section values.

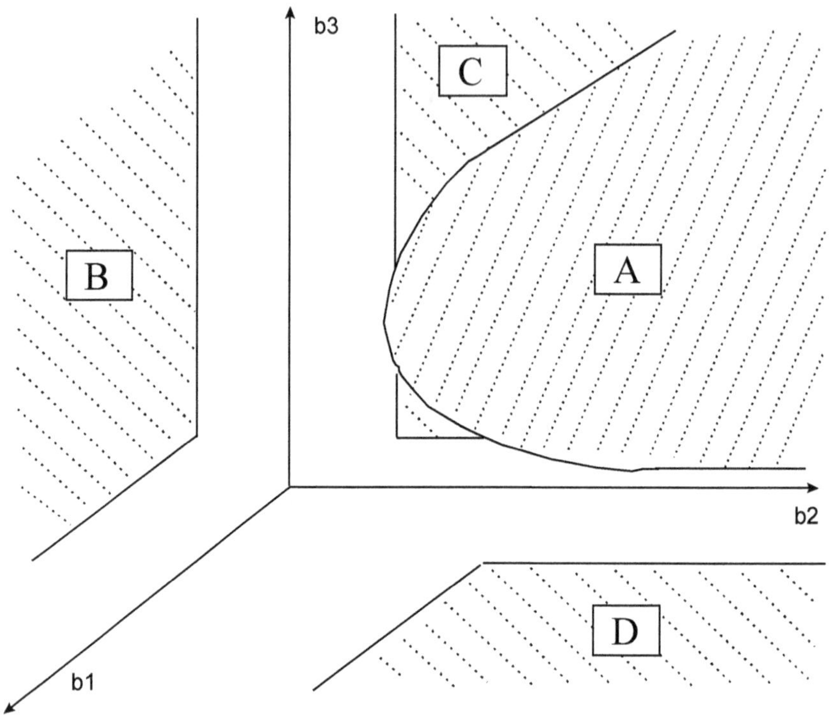

Within the main part A of the feasible region, there is a local minimum configuration at (7.038,2.079,2.784) which gives a weight of 15.969. This point is likely to be found by any search algorithm which is based on using gradients, (see section 5.7.3), and which starts from a point inside region A. A direct search, or a probabilistic search, can reveal that other local minima exist in the B,C and D regions. In B, (8.0,0,4.25) gives a weight of 17.32. In C, (0,2.829,8) gives a weight of 14.143. In D, (8.0,1.5,0) gives a weight of 12.814. The latter is thus the global minimum. The minimum weight structure is thus achieved using just the b1 and b2 bars.

In all cases, solution times are tiny compared with the other case studies.

7.2 Case Study 2: Military Aircraft Wing Flap Design: Daimler Chrysler Aerospace

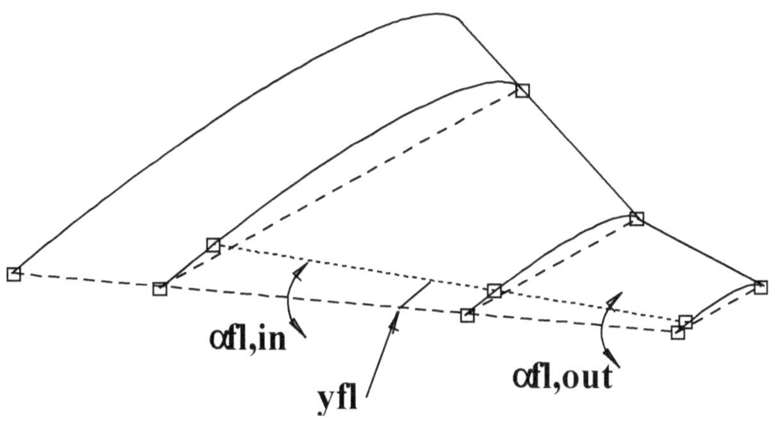

In this design case, the subject is a trailing edge flap control system on a wing which is based on DASA's X31 aircraft. An improved design is sought through introducing a split flap arrangement on the trailing edge of the wing. The idea is, if we split the flap at some point yfl, the outboard flap will obviously have a different efficiency from the inboard, due to elasticity. And we can investigate the optimal combination of flap split position, and flap deflection angles. We want the best roll rate performance of the wing, at the least cost in terms of weight of the wing. We consider a single design point, at Mach 0.9, sea level. For each design we consider, we use the HISSS Computational Fluid Dynamics panel method code for the aerodynamic analysis. The structural constraints are imposed and the mass of the aircraft is supplied using DASA's LAGRANGE structural analysis code.

Summary of case structure:

Objectives:	* Maximise roll rate
	* Minimise structural weight
Design variables	* Flap split position yfl
	* Flap settings α fl,in α fl,out
Geometric constraints	Flap split position constraints
	$1200 < yfl < 1737$ or $1817 < yfl < 2140$
	where distances are along the trailing edge.
	Flap setting constraints
	$-10 < \alpha fl,in < 20$; $-10 < fl,out < 20$
Fixed dimensions	wing surface external shape
Design evaluators	HISSSD Panel Method; LAGRANGE Structural Optimiser

Optimisation approach

A steady state multiple objective genetic algorithm was employed, using 16 generations with 16 individuals, i.e. 256 designs in all. The variables α fl,in and α fl,out were treated as continuous variables; in practice, only 3 values of the flap split position yfl were feasible due to engineering constraints so this variable was discrete.

The whole computation took approximately 2 hours on an SG Origin2000 (16 processors, 8Gb memory).

An objectives tradeoff plot of results is shown below. Further work to select a design on the tradeoff boundary was done using decision support methods referred to in section 5.8. The selected design had a more than adequate roll rate, and a 25% reduction in weight.

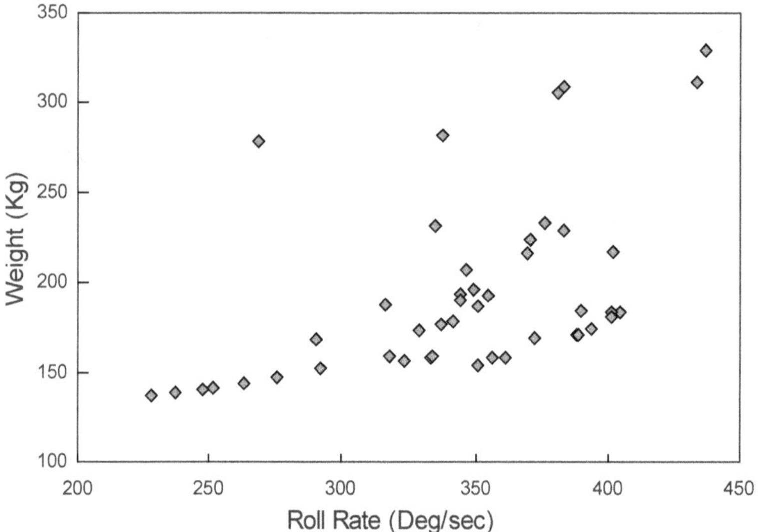

7.3 Case study 3: Military aircraft wing design: Defence Evaluation and Research Agency

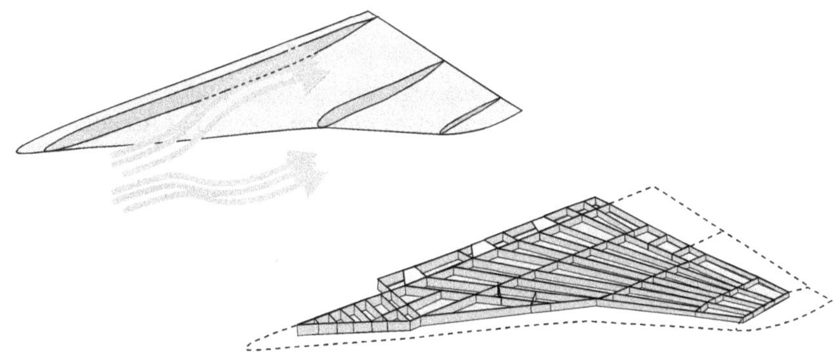

In this aircraft wing design, an existing wing on a typical modern military aircraft is being improved so as to get more fuel inside the wing and hence improve its range performance without compromising too much its manoeuvrability. The wing design can be thought of as being made up of a basic aerofoil shape which is used across the span, with the overall wing shape being varied firstly by varying the camber of the aerofoil , and secondly by varying the thickness-to-chord ratio at each of three defining stations at root, tip, and midspan positions; that is, inflating the basic aerofoil obtained after the camber has been applied. The two metrics to be traded off are the aircraft's penetration range at Mach 0.9 at sea level, and its sustained turn rate at Mach 1.4, 36000ft. For each design considered, the SAUNA CFD code was used to analyse the wing/body performance at the two design points. The mass of the aircraft was obtained through use of the STARS structural analysis code. This also says what internal volume changes inside the wing result from resizing the structural members. The two performance metrics are computed using design relations linking them with the drag, the fuel carried, the aircraft mass, and the specific fuel consumption of the engine.

Summary of case structure:

Objectives:	* Maximise supersonic performance (turn rate) * Maximise subsonic penetration range
Design variables	* Thickness-to-chord ratio at three wing stations * Camber
Constraints	'x' coordinates of points along aerofoil centre line lift constraint at Mach 0.9, sea level drag constraint at Mach 1.2, 36000 feet
Fixed dimensions	Wing span and section 'y' positions Chord length at each section

Design evaluators	SAUNA (Euler/Navier-Stokes); STARS Structural Analyser

Optimisation approach

A series of nine different spanwise thickness distributions and three different cambers were defined to provide a basic set of candidate designs. Aerodynamic analyses were conducted on these for two design points to satisfy a lift constraint at the transonic condition and a drag constraint at the supersonic condition. These, together with results from the structural design phase, provided the lift, drag and mass performance parameter data required for computing transonic range and sustained supersonic turn performance.

These results were used to define response surfaces representing range and turn rate. Based upon these, a multiobjective genetic algorithm was then used to explore the design space. An optimisation run of 10 generations, each of 32 individuals was performed. The designs identified on the Pareto boundary were then refined with the aid of decision support software and a gradient-based optimiser. The user selected a series of designs on or close to the Pareto boundary in the region of the desired STR and supplied pairwise preferences, enabling appropriate relative weightings for the objectives to be computed. The top ranked design, together with these weightings, were then transferred to a gradient-based optimiser, and a further optimisation carried out. The gradient based optimisation provided a design which is a further improvement over the other designs.

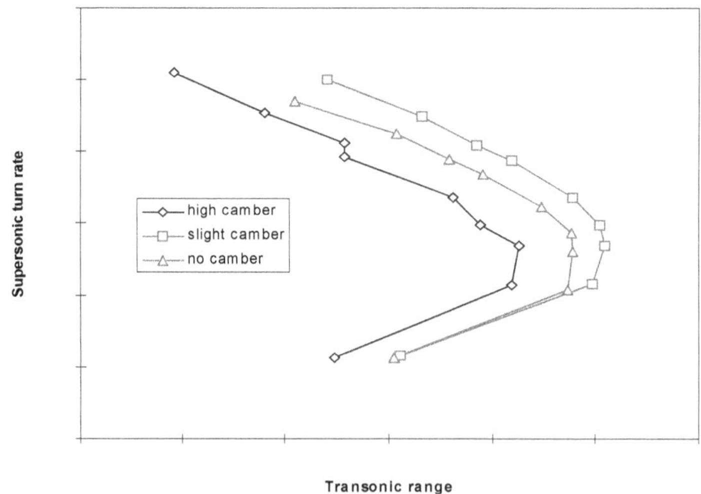

7.4 Case Study 4: Refrigerator Design: Electrolux-Zanussi

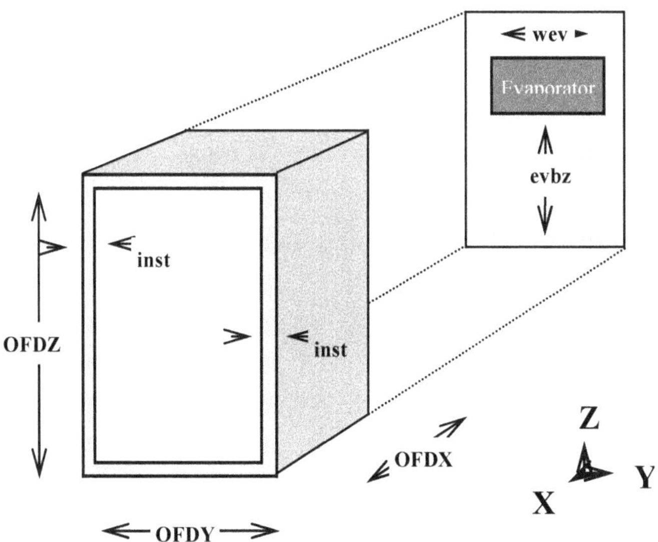

This refrigerator is a new design recently introduced by the manufacturers. The design task here is to position the cooling unit on the back of the fridge in an optimal way, and use the optimal amount of insulation material. The metrics for the optimisation are temperature evenness, the main performance measure; heat loss, which is a measure of power consumption; and area of the evaporator, to reduce cost. The STAR-CD code is used for analysis. Parametrisation has been done using the PROSTAR preprocessor, which remeshes the geometry for each design evaluation.

Summary of case structure:

Objectives:	* Minimise internal temperature gradient * Minimise heat loss * Minimise area of the evaporator
Design variables	* Height of evaporator bottom * Evaporator depth * Evaporator thickness * Evaporator width * Insulation thickness
Geometric constraints	Each parameter has a minimum and a maximum bound
Fixed dimensions	Outside dimensions of the fridge Size of the condenser
Design evaluators	STAR-CD CFD/Heat Transfer Commercial Code

Optimisation approach

A multiple objective genetic algorithm was employed. Each design analysis used a 50000 cell model, and simulated a 30 minute transient period. One run took 1.6 hours on a SG Origin2000. 4 generations of 16 individuals were run. These took around 5 days computing time. Of the 64 designs analysed, 7 were Pareto-optimal with respect to the others. Of these, 4 were identified as being candidates for 'best overall design'. A 'parallel coordinates' plot of these 4 solutions is shown below. Solution 10 minimises both temperature deviation and heat dispersion, so it is preferred even though it is worse in terms of evaporator extension, which was considered lower priority.

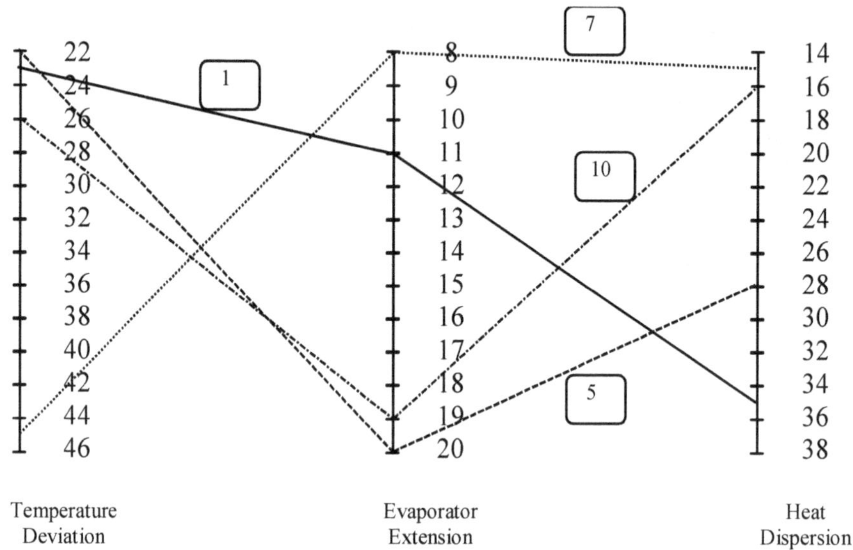

7.5 Case Study 5: Boiler Design: Calortecnica

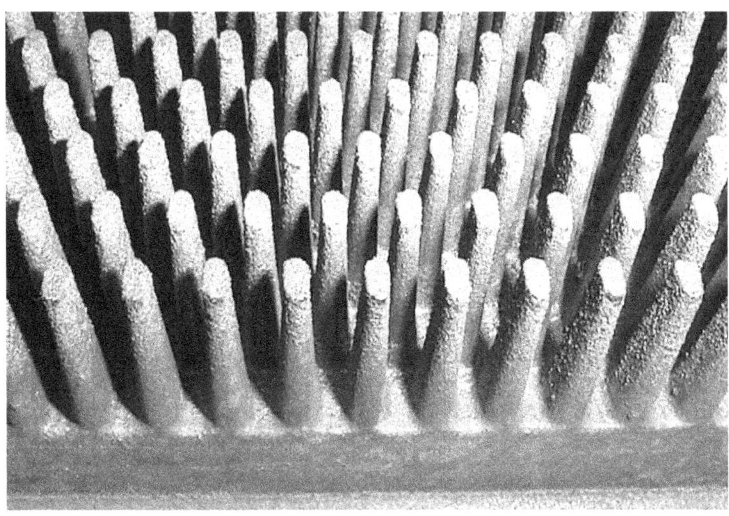

This design case involves a heat exchanger which is a main component of an industrial gas-fired boiler and is the main product line of the company concerned. The gas flows across the fins of the heat exchanger were modelled using the FIDAP CFD code. In the first instance a single column of fins was considered, at various Reynolds numbers. From these, the performance of the complete heat exchanger was constructed. In optimisation terms the shape of the fins (parameterised using Bezier curves) is being changed in order to maximise the performance of the unit, by maximising heat flux per unit area, whilst also seeking to minimise pressure drop across the heat exchanger. The gas needs to be drawn through the heat exchanger using a fan, and the smaller the pressure drop, the smaller the fan needed. Hence the smaller the production cost.

Summary of case structure:

Objectives:	* Maximise convective heat flux * Minimise pressure loss
Design variables	* Control points of the Bezier curves
Geometric constraints	Bounds on control point variation Minimum radius of curvature of fins is 1.5mm
Fixed dimensions	Overall area of heat exchanger
Design evaluators	FIDAP (Commercial code for CFD/Heat Analysis)

Optimisation approach

A multiple objective genetic algorithm was employed, using 16 generations with 16 individuals, i.e. 256 designs in all. Selected designs on or near the Pareto boundary, as shown below, were used in decision support analysis to quantify relative weights of the objectives. Subsequently, a composite objective function was optimised to produce a final best design, which is highlighted in red. The new design provided a 22% reduction in weight and a 35% reduction in volume compared with the old production model.

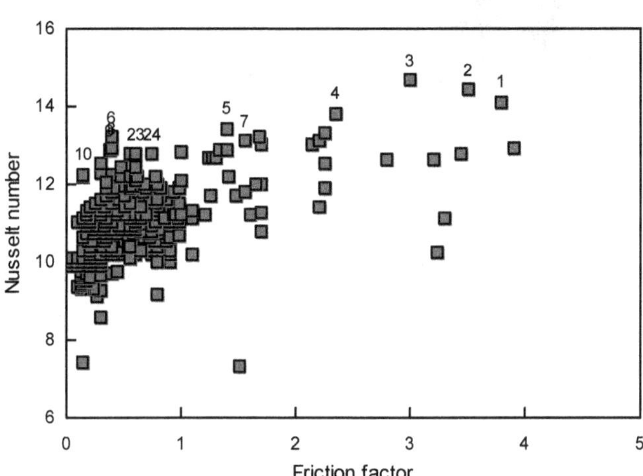

7.6 Case Study 6: Cantilever Beam Topology Optimisation: Univ. Of Sydney

The cantilever beam structure shown below is to be optimised. The structure is made of isotropic material and is subject to plane stress through a single static load The initial candidate region is 100mm x 50mm x 1mm. The left hand edge of the beam is fixed, and the rest of the structure is only allowed to move vertically. The applied load is an upwards shear force of 20 Mpa's acting over a distance of 10mm centrally located on the right hand edge. The objective is to equalise the strain energy density throughout the structure. The domain is subdivided into a 60x30 mesh.

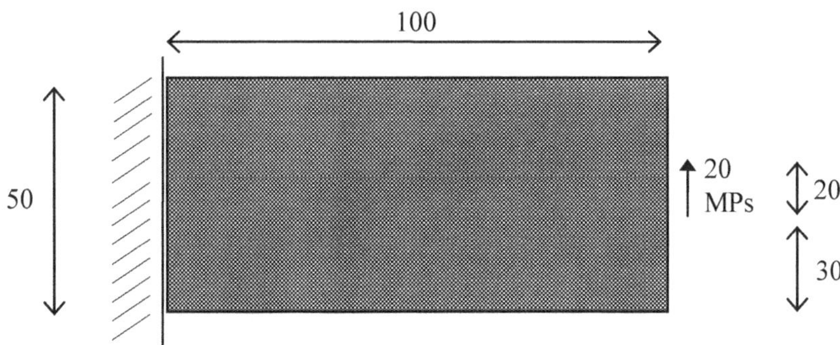

Summary of case structure:

Objective:	* minimise maximum deviation of element strain energy density from maximum strain energy density
Design variables:	* the 1800 integer variables, valued (0,1), which determine whether material is present (1) or absent (0) in the mesh element concerned
Geometric constraints:	* fixed 100mm x 50mm material domain * fixed left hand edge
Design evaluator	finite element analysis code within EVOLVE optimisation software; using linear quadrilateral elements

Optimisation approach

The search strategy adopted is based on starting from an over designed specimen. The single static loading case is analysed to find the strain energy value in each element. Material is then removed from regions which have very low strain energy density values. Following removal, the design is reevaluated to produce updated element strain energies, which are in turn used to decide further removals. The strain energy threshold used to trigger removals is progressively increased as the iterations proceed. The iteration is stopped when all strain energies are within a specified margin of the maximum. The final topology reached is shown below. The final volume is 42.3% of the initial volume. The solution time was 2.8 hours using a single Pentium90 processor.

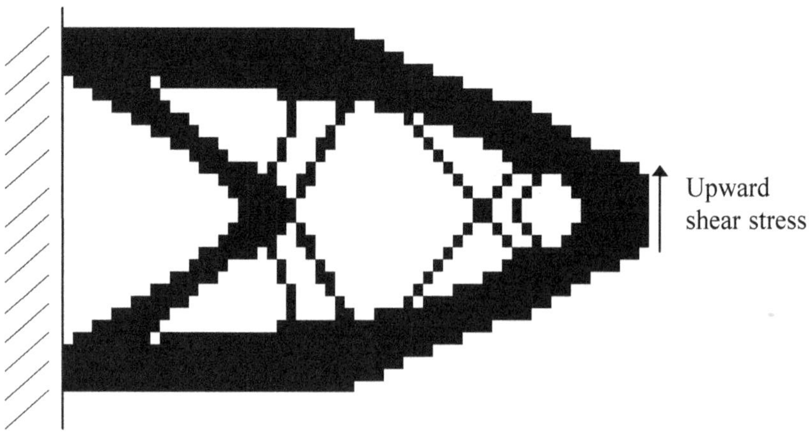

7.7 Case Study 7: Casting Process Design: Enginsoft Trading

This case study is from manufacturing process design and concerns a casting process. The objective is to cast the component shown below. Modelling the casting process involves modelling the flow of molten metal into the mould, modelling the cooling which takes place, and modelling the state change from liquid to solid, together with the creation of air inclusions in the cast. This produces a coupled set of Navier Stokes equations, Fourier heat equations, and state equations. The aim here is to determine the best geometrical arrangement and starting conditions for the casting process, and also the best mould characteristics. There are three objectives of interest. Two are measures of the quality of the product being cast, these being its hardness and its porosity. The other is the weight of the casting before it is trimmed to its final shape. Weight is really a measure of how much excess material is trimmed off and therefore wasted, and hence is a measure of cost.

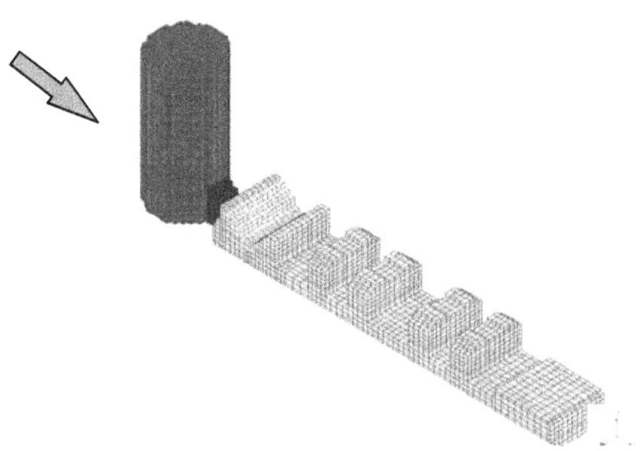

Summary of case structure:

Objectives:
* Maximise hardness
* Minimise weight
* Minimise porosity

Design variables
* Initial cast temperature
* Heat transfer coefficients between cast and mould
* Feeder height
* Feeder diameter
* Feeder neck area

Constraints	Range limits on all variables
Fixed dimensions	Geometry of the part being cast
Design evaluators	MAGMASOFT (Navier-Stokes flow, Fourier heat transfer, moving boundary)

Optimisation approach

A first stage was done using a standard multiobjective genetic algorithm, using 4 generations of 16 individuals. Each design evaluation took 20 minutes on a HP C200, so total time was around 21 hours. Because the evaluation is relatively expensive, a second stage was done using a neural net approximation based on the first stage results. A further run of 8 generations of 16 individuals was then carried out, and a Pareto set of 11 designs was extracted from these. A decision support exercise was then carried out to rank the designs and to understand and verify the consistency of the design choices made. The best 3 results from this were then checked by running full simulations. These are compared below to the baseline production design.

	Variables					Design Objectives		
	T_{init}	HT	H_{fd}	D_{fd}	A_{n}	Hardness	Casting weight	Porosity
New designs	1352	460	105	33	400	217	3.28	2.34
	1365	400	89	32	341	218	3.11	1.80
	1336	400	112	31	400	218	3.24	3.30
Baseline design	1360	700	115	55	248	207	4.53	1.27

The best design identified showed significant improvements in hardness and weight, at the cost of a slight increase in porosity.

8 Future Prospects For Optimisation

At the present time we are a long way from being able to tackle large scale engineering design optimisation tasks routinely. The case studies outlined in section 7 give an indication of the type and size of problems which are being tackled today. But for the future, there need to be capability improvements in a number of areas to enable optimisation to be used as an integral part of mainstream design.

The limiting factors have mostly been referred to already. They concern

- how long it takes to perform one design evaluation including assembling and transferring data,
- how many design evaluations are needed in a search, and
- limitations on whether a design evaluation can be done at all imposed by modelling methods and evaluation toolsets.

- *Computing power*

Processor power has increased steadily by more than 20% per year for the last 30 years and is set to continue for at least the next decade or two. Memory and disk space availability have also increased , though memory speeds are not growing as fast as CPU speed. In that period, improvements in analysis algorithms have resulted in an even bigger speedup than the hardware. In the next decade we can expect a 5000-fold increase in speed of top-of-the-range machines.

For optimisation, network speeds are significant in cases where widely dispersed machines need to access the same updated CAD model. Current telephone modems operate up to 56Kbits/sec. The fastest LANs on the market can work at up to 1 Gbits/sec running TCP/IP. Fibre optics technologies (Wave Division Multiplexing, Dense Wave Division Multiplexing) will take speeds into the 40 Gbits/sec and eventually to the 10 Tbits/sec range via DWDM, aiming towards 75 Tbits/sec fibre capacity. Using the technology available in the case studies discussed in section 7, transferring 1Mbyte of CAD data over a WAN of, say, 4 Mbps could take as much as 1 hour. Speeds such as quoted could allow this transfer to take place in 1 second.

One further commercial aspect which is significant for optimisation concerns the licensing of application codes. Because parallel design evaluation is of particular relevance to optimisation, arrangements for multiple licenses for application codes are potentially a major limiting factor. This needs to be foreseen by prospective optimisation users, and dealt with through negotiation with the code providers.

- *Search*

Approaches to conducting design searches have been outlined briefly in section 5.7. Much development is still being done on advanced search method to exploit and integrate designers' domain knowledge of problem and solution methods, through rule based and other methods. Progress in realising this research is steady but there is still a long way to go.

- *Toolsets and modelling*

As alluded to in section 6.3, it is still common to find that some areas of engineering are inconveniently difficult to model and analyse. Such difficulties are very specific to the field concerned, and in some instances are a major barrier to progress.

Except in some areas of 'Toolsets and modelling' there is no reason to suppose that there are major barriers to future optimisation progress. The scale of problems which can be addressed will move steadily towards complete designs of the most complex engineering products. We can look forward to an unimpeded and relatively rapid growth in optimisation capabilities.

9 References

[1] *Competitive strategy: techniques for analysing industries and competitors.* Michael E. Porter. The Free Press. 1980

[2] *Competitive advantage: creating and sustaining superior performance.* Michael E. Porter. The Free Press. 1998

[3] *On competition.* Michael E. Porter. Harvard Business Review book series. 1998

[4] *Competing for the Future.* Gary Hamel , C.K. Prahalad. Harvard Business School Press. Boston. 1994

[5] http://www.sgi.co.nz/success_teamnz.html

[6] *The relationships between data, information and knowledge based on a preliminary study of engineering designers.* S. Ahmed, L.T.M. Blessing, K.M.Wallace. DETC 99, ASME Design Theory and Methodology, Las Vegas, Nevada, USA. 1999

[7] *FRONTIER: Open System for Collaborative Design using Pareto frontiers.* D. Spicer. R.Ae.S. Conference on Multidisciplinary Design and Optimisation. London. Oct. 1998

[8] *JAVA language*
www.javasoft.com/doc/language_environment

[9] *The Essential CORBA: Systems integration using distributed objects.* T.J.Mowbray & R. Zahavi. 1995. J. Wiley & Sons.

[10] *Numerical Methods for Unconstrained Optimisation.* W. Murray (ed). 1972. Academic Press

[11] *Genetic algorithms in search, optimisation and machine learning.* D.E. Goldberg. 1988. Addison-Wesley

[12] *Constraint Programming: in pursuit of the Holy Grail.* R.Bartak. Proc. WDS99. Prague. June 1999.

[13] *Implementation and relative efficiency of quasirandom sequence generators,* B.L. Fox, ACM Transactions on Mathematical Software Vol. 12, No. 4, Dec 1986.

[14] *The design and analysis of industrial experiments.* O.L. Davies. 1963. Oliver & Boyd.

[15] *Taguchi techniques for quality engineering.* P.J. Ross. 1996. McGraw-Hill.

[16] *Numerical Methods for Constrained Optimisation.* P.E. Gill & W. Murray (eds) 1974 Academic Press.

[17] *Fortran codes for mathematical programming: linear, quadratic and discrete.* A.H. Land & S. Powell. 1973 Wiley-Interscience

[18] *Nonlinear Multiobjective Optimization* K. Miettinen Kluwer Academic Publishers, Boston, 1999

[19] *Multiple criteria optimization* R.E. Steuer. J. Wiley & Sons. 1986

[20] *Multiple objective decision making - methods and applications* C.L. Hwang & A.S.M. Masud. Springer Verlag. 1979.

[21] *Hybridisation of a multi objective genetic algorithm, a neural network and a classical optimiser for a complex design problem in fluid dynamics.* C. Poloni, A. Giurgevich, L. Onesti & V. Pediroda. CMAME Journal. Dec 1998

[22] *Gaussian processes for regression.* C.K.I. Williams & C.E. Rasmussen. Advances in Neural Information Processing Systems 8. MIT Press, 1996

[23] *Interdigitation: a hybrid technique for engineering design optimisation employing genetic algorithms, expert systems, and numerical optimisation.* D.J. Powell, M.M. Skolnick & S.S. Tong. In *Handbook of genetic algorithms.* L. Davis (ed) Van Nostrand Reinhold. 1991.

[24] *Parallel coordinates: a tool for visualising multidimensional geometry.* A. Inselberg & B. Dimsdale. 1990 Proc. IEEE Conf. Visualisation '90. pp361-378

[25] *Applied Optimal Design.* E.J. Haug & J.S. Arora. John Wiley & Sons. 1979

[26] *Multiobjective aeroelastic optimisation.* M. Stettner & W. Haase. . NATO RTO-AVT Symposium on Aerodynamic Design and Optimisation of Flight Vehicles in a Concurrent Multidisciplinary Environment. Ottawa. Oct 1999

[27] *The application of Pareto frontier methods in the multidisciplinary wing design of a generic modern military delta aircraft.* S.V. Fenwick & J. Harris.

NATO RTO-AVT Symposium on Aerodynamic Design and Optimisation of Flight Vehicles in a Concurrent Multidisciplinary Environment. Ottawa. Oct 1999

[28] *Trial report definition: ZANUSSI test cases.* P. Gozzi, F. Noviello, E. Gellner. ESPRIT project 20082 FRONTIER Deliverable 12.2. Feb 1999

[29] *Report on initial experiences with Pareto frontier system.* S.R. Stella & L. Marcuzzi. ESPRIT project 20082 FRONTIER Deliverable 11.2 Jan 1997

[30] *Evolutionary Structural Optimisation* Y.M. Xie & G.P. Steven Springer, 1997

[31] *Application of multiobjective genetic algorithm to the simulation of casting proesses.* EnginSoft Trading s.r.l. June 1998

www.ingramcontent.com/pod-product-compliance
Lightning Source LLC
Chambersburg PA
CBHW060507220326
41598CB00025B/3584
9781910643273